Laser Beam Welding

Tobias Buccho

Kelsiehanson Verlag

DEDICATION

To My Wonderful Readers

CONTENTS

ACKNOWLEDGMENTS

Thank you to everyone who strives to grow and help others grow

1 ABSTRACT

This book deals with pulsed mode laser beam welding (LBW) on 2 mm thick Ti-6Al-4V alloy sheets. ANSYS Fluent 16.0 version is utilized to develop a generic model with Vc++ code. The model is validated with test results. The optimal LBW parameters are obtained bymaximizing the depth of penetration (DP) and minimizing the weld bead width (BW). To minimise the number of numerical simulations, Taguchi approach is adopted.The LBW parameters considered are pulse frequency, pulse width and welding speed,theabsorbed power is kept as constant.The performance indicators are width of fusion zone, depth of penetration, heat affected zone and maximum temperature. The optimal LBW parameters are found to be dependent onthe process indicators. A simple reliable modified Taguchi approach is followed to get a set of optimal LBW parameters for maximum DP and minimum BW.

Index Terms—Absorber power; Bead width; Computational Fluid Dynamics (CFD); Depth of penetration; Laser beam welding (LBW); Pulse frequency; Pulse width; Ti-6Al-4V alloy.

2 INTRODUCTION

Laser beam welding (LBW) is an efficient welding method in industrial manufacturing. Because of high density and low heat input to the weld, it offers low distortion in components. Titanium and its alloys have low density, good corrosion resistance, and high operating temperature. They are being used extensively in aerospace, automotive, biomedical and nuclear industries [1, 2]. However, it has difficulty in conventional machining and welding. Pulse wave (PW) and continue wave (CW) lasers are introduced to overcome such problems. The welding parameters in CW laser are: (i) laser power; (ii) welding speed, and (iii) beam diameter.PW laser considers energy and pulse duration as additional factors [3-7]. Kumar et al. [3] have performed Nd:YAG laser welding on Monel 400 sheets to examine the effect of pulse energy. They have noticed enhancement in peak temperature, weld width and the heat affected zone (HAZ) size with increasing pulse energy. Cheepu et al. [8] and Muralimohan et al. [9] have examined the possibility of improving properties of dissimilar material friction welds. Yang et al.[10] have developed a 3-D model for a moving Gaussian laser heat source. Akman et al. [11] have made investigations on the weld quality of Ti-6Al-4V alloy. They have used pulse Nd:YAG laser and controlled pulse duration and pulse energy. Mueller et al. [12] have determined the weld bead geometry of Ti-6Al-4V alloy specifying the welding speed. Welding with high speed has the tendency of gas entrapment. Pulsed Nd:YAG laser has several advantages in providing narrow weld bead and HAZ in addition to lowering of the heat input, deformation and solidification time. PW lasers are well suited for welding thin sheets [13-16].

Kong et al. [17] have developed a double-ellipsoidal volume heat source model. They have made simulations on the GMAW (gas metal arc welding) of ASTM A36 carbon steel. The laser beam is simulated considering a cylindrical volume heat source model. Frewin and Scott [18] have considered unsteady three-dimensional heat flow during pulsed Nd:YAG laser welding. The convective flows in the melt pool are neglected. Gaussian energy distribution is assumed while estimating the transient temperature cross-sections along with the dimensions of the fusion and HAZ. Their estimates are close to the measured data. Satyanarayana et al. [19] have proposed a generic model using ANSYS Fluent 16.0 version with VC++ code. Their model is validated considering test data of several materials and welding process. Taguchi approach provides the %Contribution of LBW parameters on the performance indicators through design of few experiments and performing ANOVA (analysis of variance) [20–28]. Kumar et al. [29] have made LBW on 1mm thick Ti-6Al-4V alloy sheets varying laser power and welding speed. Numerical simulations are comparable with test results. Tensile strength of samples annealed at 9800C is 1048 MPa. They have achieved 3 to 4% improvement on the tensile strength of the conventional welded samples. Jiang et al. [30] have performed numerical simulations and experiments on Nd:YAG laser welding of Ti-6Al-4V varying average power, beam diameter and pulse energy. Modified Taguchi approach is considered to identify the optimal parameters for machining and LBW of titanium and zirconium alloys [31-34].

Kumar et al. [35] have examined the effect of LBW parameters on butt joints. Rajulu et al. [36] have studied the weld bead characteristics by conducting experiments varying the pulse frequency, pulse width, welding speed, and pulse energy. Caiazzo et al. [37] have presented optimal welding processing conditions minimizing undercut and porosity. Liu et al. [38] have examined the effect of heat input on the tensile damage evolution of pulsed Nd:YAG laser welding of Ti-6Al-4V alloy. Cai et al. [39] have utilized Ti filler with laser to improve the joint strength. Junaid et al. [40] have reported that laser peak power influences the fusion zone oxygen contents and grain size, whereas peak power and heat input control the weld pool shape. Kashaev et al. [41] have examined the effect of Nd:YAG LBW on butt and T-joints. Hong et al. [42] have made interesting studies using IPG YLR-1000 fiber laser. Fan et al. [43] have made a comparative study on microstructures and mechanical properties using Nd:YAG, diode and fibre lasers. Boccarusso et al. [44] have utilized LBW and examined the influence of weld bead morphology on Ti-6Al-4V alloy under fatigue.

This paper incorporates the Taguchi approach in the numerical simulations of thin Ti-6Al-4V alloy sheets to obtain the optimal LBW process parameters. Initially numerical simulations are validated through a test data. Narrow weld bead width with full depth of penetration is arrived in numerical simulations varying the LBW parameters (viz., pulse frequency, pulse width and welding speed).Taguchi L9 orthogonal array is chosen to minimize the number of numerical simulations and develop empirical relations for the performance indicators (namely, weld bead width, depth of penetration, heat affected zone, and maximum temperature) in terms of LBW parameters. A multi-objective optimization procedure is adopted for finalizing a set of optimal LBW parameters.

3 MATERIALS AND METHODS

1. *Thermo-fluid Analysis*

Thermal history in the weld is required to assess the weld joint strength. Optimal LBW parameters are to be identified to improve the weld quality. Trial and error methods are time consuming and expensive. Computational welding mechanics (CWM) is playing an important role in welding science and Technology. A CFD Model for thermo-fluid analysis is required for generating the weld bead profile. This section highlights the basic equations that are been solved through User Defined Functions (UDF) with the code relevant to finite volume method in ANSYS. The Navier-Stokes (N-S) equations and k-ε equation in mushy zone and weld pool regions are considered as in [32]

The continuity equation is

$$\frac{\partial \rho}{\partial t} + \nabla \cdot \left(\rho \vec{V}\right) = 0 \qquad [1]$$

Here ρ is the density, t istime, and $\vec{V}$ is the weld pool velocity.

Variation of density (ϱ) with in solidus temperature (T_s) of 1878 K is

$$\varrho = 4466.2 - 0.154\ T \qquad [2]$$

Variation of ϱ *above* T_s *and below the liquidus temperature of 1928 K is*

$\varrho = 31.34\ T - 18467$ [3]

and above the liquidus temperature is

$\varrho = 5076.8 - 0.68\ T$ [4]

Specific heat of solid is 670 J/kg K, in case of liquids it is 730 J/kg K and viscosity, Melting heat are as follows 0.005 Kg/m.s 370000 J/kg.

The momentum equation is [45]

$$\frac{\partial(\rho\vec{V})}{\partial t} + \nabla \cdot (\rho\vec{V}\vec{V}) = -\nabla p + \mu\nabla^2\vec{V} + \rho\vec{g} + \vec{S}_W$$ [5]

Here, p *is the static pressure,* μ *is the viscosity and* $\rho\vec{g}$ *is the gravitational body force. Momentum frictional dissipation or the source term in the mushy zone is considered as in [46, 47].* $f_L = 0$*, within the solidus temperature.*

$\overrightarrow{S_w} = 10^3 \vec{w}\ A_{mush}$ [6]

$f_L = 1$*, above the liquidus temperature*

$\overrightarrow{S_w} = 0$ [7]

In between liquidus and solidus temperatures,

f_L*=0.02T-37.56* [8]

$\overrightarrow{S_w}$ *=((0.02T-37.56)*^{3} *+ 10*$^{-3}$*)*$^{-1}$ *(1-(0.02T-37.56))*^{2}A_{mush} [9]

Here A_{mush} *is the mushy zone constant, and* $\vec{w}$ *is the pull velocity.*

The enthalpy-porosity technique [48] is adopted while tracing the interface resulting from the phase change.

The energy equation is

$$\frac{\partial(\rho H)}{\partial t} + \nabla \cdot (\rho \vec{V} H) = \nabla \cdot (\kappa \nabla T) + S \quad [10]$$

Here $H(\equiv h + \Delta H)$ *is the enthalpy;* $h = h_{ref} + \int_{T_{ref}}^{T} C_P \, dT$ *is the sensible heat;* C_P *is the specific heat;* h_{ref} *is the reference enthalpy;* κ *is the thermal conductivity; and* S *is volumetric heat source.*

The latent heat,

$\Delta H = 0$ *for a solid* *and* $\Delta H = m_L L$ *for liquid* [11]

Here L is the latent heat of the liquid metal. The velocity normal to top surface of the flat weld pool is zero.

The shear stress,

$$\tau = \frac{\partial \sigma}{\partial T} \nabla T \qquad [12]$$

The convection and radiation boundary conditions on all surfaces are considered. In addition, on the top surface, a transient heat flux (which is produced by the beam laser) is [49, 50]

$$\kappa \frac{\partial T}{\partial z} = q - h_c (T - T_a) - \sigma_{SB}\, \varepsilon \left(T^4 - T_a^4\right) \qquad [13]$$

Here $h_c = 2.4 \times 10^{-3} \varepsilon T^{1.61}$; ε *=0.708±0.012;* σ_{SB} *= 5.67108 W/m² K⁴; and* T_a *is the ambient temperature.*

For laser, the Gaussian distribution of volumetric heat flux (q_v) is

$$q_v = \frac{2P}{\pi r_0^2 \delta} i(t) \exp\left(1 - \frac{r^2}{r_0^2}\right)\left(1 - \frac{y}{\delta}\right) \qquad [14]$$

*Here **P** is the absorbed laser power; r_0 is the laser beam radius; δ is the total thickness; r is the current radius (i.e., distance from the cone axis); and y is the current depth. In the Pulsed laser welding, the time-based function i(t) =1, (i.e., the pulse is active), whereas i(t) =0, when the pulse is inactive.*

Regarding welding process simulation, the computational domain and the geometry of component is created with fine mesh close to the heat source. Coarse mesh is created away from the heat source. Appropriate phases and turbulence model are selected. User defined codes are introduced. Thermal and physical properties are specified, and the boundary conditions are applied. The number of time steps and iterations are specified for solution convergence. Laser beam activation is done with starting and end positions. Based on the temperature distribution the weld bead profile is generated. Residual stress distribution can be obtained by exporting the geometry and solution file from fluent to FEA solver.

2. *Validation of Numerical Simulation*

Experiments are conducted to validate the temperature dependent properties, Modelling and Numerical simulation of Pulsed Laser beam welding of 50 x 20 x 2 mm Ti-6Al-4v alloy plates. Using a 16kw Nd:YAG pulsed mode with CNC station the LBW process simulation is performed to generate weld pool cross section, for the Pulse frequency-14Hz, Pulsewidth-8ms, welding speed 1.5 mm/sec and absorber power 350 W. Fig.1 shows the weld bead profiles of experimental and numerical CFD simulations. Numerical simulation provides 1.102mm weld width and 1.204 mm weld depth. Experiment results indicate 1.064mm width of fusion zone and 1.286mm weld depth. Numerical simulation results indicate the validity of the test data. This study confirms the validity of temperature dependent properties of Ti-6Al-4V alloy. It also confirms the adequacy of the model, numerical simulations and weld pool shape of test run is as shown in figure 2.

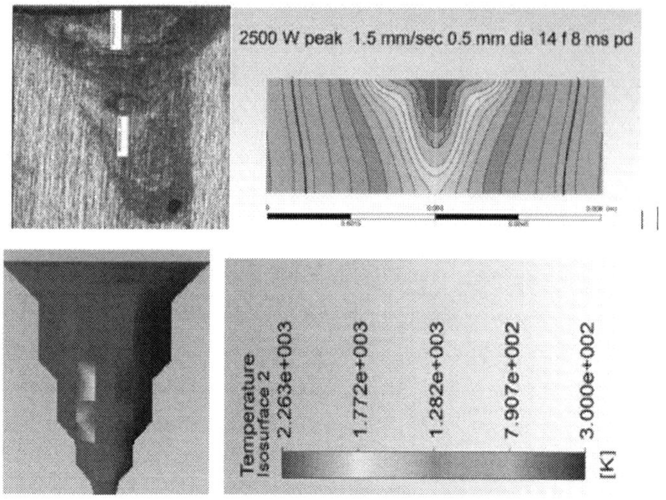

Figure-1: Weld bead through experiment and simulated temperature contour plot

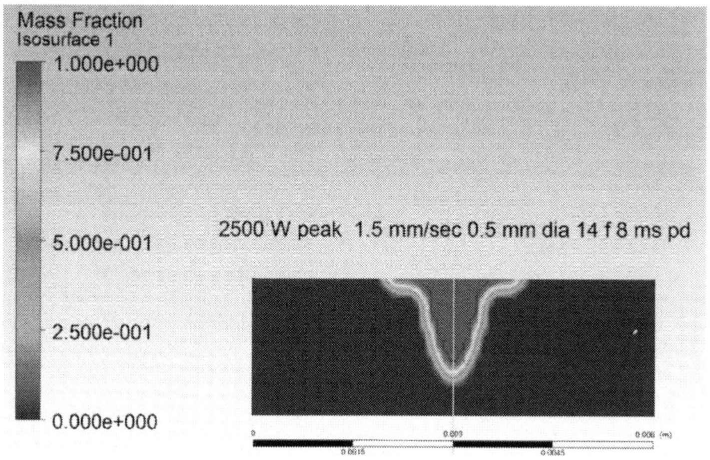

Figure-2: Weld pool shape for Test Run

3. Modified Taguchi Approach

Taguchi method [28, 51] recommends a suitable OA (orthogonal array) for the intended number of factors or process parameters (n_p) *and the levels* (n_l) *assigned to each factor to conduct few experiments and obtain optimal parameters. The minimum test runs (**N**$_{Taguchi}$) required is [28]*

$$N_{Taguchi} = 1 + n_p \times (n_l - 1) \quad [15]$$

For $n_p = 3$ *and* $n_l = 3$, ***N**$_{Taguchi}$ = 7, Taguchi method [28] recommends* L_9 *orthogonal array (OA). Table-1 presents numerical simulations of the performance indicators (viz., width of fusion zone, α (mm), depth of weld, β (mm), heat affected zone, γ (mm) and maximum temperature, δ (K)) including the simulated weld bead profiles for the nine test runs. For* L_9 *OA, substitution of the number of test runs, **N**$_{Taguchi}$=9 and the levels,* $n_l =$ *3 in equation (15), one can find* $n_p = 4$. *That means 4 factors can be accommodated in 9 test runs. As in [52-63], without loss of generality, an additional or fictitious parameter (D) is considered in Table-1. Table-2 presents analysis of variance (ANOVA) results. It is seen that %contribution of C is significant on the grand mean value of both α and β. %Contribution of A, B and C on α are 17.6, 33.9 and 47.6% respectively, whereas 8.1,33.92 and 56% are respectively on β. Sum of the %Contributions of A, B and C on α and β are 99.1 and 98%. Hence, the %Contribution of D on both α and β are 0.9 and 2%, which are nothing but the error (%).*

From ANOVA Table-2, the optimal LBW parameters to achieve minimum α (width of fusion zone) are $A_1B_1C_3$, wherein subscripts denote the levels of the parameters. The optimal LBW parameters to achieve maximum β (depth of weld) are $A_3B_3C_1$. The above two sets of optimal LBW parameters to achieve minimum α and maximum β are different. Since, these two sets of LBW parameters are not available in L_9 OA of Table-1. Confirmation tests for these two sets of LBW parameters to obtain the performance indicators are mandatory.

Additive law [28] provides the estimates of width of fusion zone and depth of weld from the mean values of ANOVA Table-2. The evaluation procedure for the estimates of the performance indicators is explained below. Let ψ be the performance indicator and $\widehat{\psi}$ is its estimate for the LBW parameters (A_i, B_j, C_k, D_l) varying the subscripts i, j, k, l from 1 to 3. Designating $\psi(A_i)$, $\psi(B_j)$, $\psi(C_k)$ and $\psi(D_l)$ as mean values of ψ corresponding to the ith level of A, j^{th} level of B, k^{th} level of C and l^{th} level of D. The grand mean of ψ is designated by ψ_{mean} for the 9 test runs. Following the additive law, estimate $\widehat{\psi}$ for the specified (A_i, B_j, C_k, D_l) is [31]:

$$\widehat{\psi} = \psi(A_i, B_j, C_k, D_l) = \psi(A_i) + \psi(B_j) + \psi(C_k) + \psi(D_l) - 3 \times \psi_{mean} \qquad [16]$$

For the specified (A_i, B_j, C_k), equation (16) reduces to [31]

$$\widehat{\psi} = \psi(A_i, B_j, C_k) = \psi(A_i) + \psi(B_j) + \psi(C_k) - 2 \times \psi_{mean} \qquad [17]$$

Estimates from equations (16) and (17) provide the different ψ values, whose deviation is $\psi(D_l)-\psi_{mean}$. Three deviations can be found for the three levels (l=1, 2, 3). As in [31], superimposing the minimum and maximum deviation values to the estimates using equation (17), the range of estimates is arrived. Table-3 gives estimates of α, β, γ and δ. Numerical simulations are within the estimated range. Tables 4 to 7 give the estimates of α, β, γ and δ for all possible 27 combinations $((((A_i B_j C_k), k = 1,2,3), j = 1,2,3), i = 1,2,3)$ of LBW parameters.

Empirical relations are developed for α, β, γ and δ in terms of A, B, and C using the additive law from the mean values of ANOVA Table-2 in the form

$$\alpha = 1.77 + 0.551\xi_1 + 0.024\xi_1^2 + 0.763\xi_2 - 0.094\xi_2^2 - 0.888\xi_3 + 0.3\xi_3^2 \quad [18]$$

$$\beta = 1.39 + 0.128\xi_1 + 0.11\xi_1^2 + 0.292\xi_2 + 0.006\xi_2^2 + 0.375\xi_3 + 0.004\xi_3^2 \quad [19]$$

$$\gamma = 1.99 + 0.065\xi_1 - 0.049\xi_1^2 - 0.078\xi_2 - 0.004\xi_2^2 - 0.196\xi_3 + 0.005\xi_3^2 \quad [20]$$

$$\delta = 2319.11 - 128.17\xi_1 + 49.17\xi_1^2 + 124.33\xi_2 + 16.67\xi_2^2 - 46.17\xi_3 + 74.17\xi_3^2 \quad [21]$$

Here, $\xi_1 = 0.5A - 7;\ \xi_2 = B - 8;\ and\ \xi_3 = 2C - 3$.

4. Multi-objective optimization

To achieve minimum bead width (α) and maximum depth of penetration (β), two different sets of LBW parameters identified are ($A_1B_1C_3$) and ($A_3B_3C_1$). To select a set of LBW parameters, a simple and reliable multi-objective optimization technique [56, 57] is considered. In this technique, a single objective function (ζ) is constructed as a function of α and β after normalizing with their maximum values α_{max} = 4.34 mm and β_{max} =2.378mm. Introducing the weighing factors $\omega_1\,(\geq 0)$ and $\omega_2\,(\geq 0)$, which satisfy $\omega_1 + \omega_2 = 1$, the single objective function (ζ) is constructed in the form [56, 57]

$$\zeta = \omega_1\,\zeta_1 + \omega_2\,\zeta_2 = \omega_1\left(\frac{\alpha}{\alpha_{max}}\right) + \omega_2\left(\frac{\beta_{max}}{\beta} - 1\right) \qquad [22]$$

Minimization of ζ provides the maximum β (depth of penetration) and minimum α (bead width) for a set of LBW parameters. For common optimum process conditions, $\omega_1 = \omega_2 = \frac{1}{2}$ (equal weighing) is considered in Table-8 and performed ANOVA on values of ζ. The set of optimal LBW parameters for the minimum ζ selected are $A_1B_3C_1$. They correspond to 12 Hz pulse frequency, 9ms pulse width and 1 mm/sec welding speed, respectively. If fact, the identified set is not available in Taguchi L_9OA of Table-1. Numerical simulations are performed by specifying the identified process parameters ($A_1B_3C_1$) and obtained the weld bead profile. Table-9 gives the details of the weld bead profiles corresponding to the single and multiple objective optimization problems. Numerical simulations are within the estimated range from the developed empirical relations.

3. Results

Numerical simulations of pulsed mode LBW on 2mm thick Ti-6Al-4V alloy sheets performed and validated with test results. To arrive optimal LBW parameters, the performance indicators (namely, width of fusion zone and depth of penetration) are estimated for the specified input variables in each test run of the Taguchi L_9 orthogonal array. Empirical relations of the performance indicators are developed in terms of input variables considering the mean values and additive law. Modified Taguchi approach provides the range of estimates for the performance indicators. Numerical simulations for the specified input variables are within the range of estimates. Incorporation of Taguchi approach in numerical simulations reduces the number of trails runs thereby reduction in computations time in identifying the optimal LBW parameters.

Table- 1: Numerical simulations on LBW of Ti-6Al-4V alloy.

(a) Levels of LBW parameters

LBW parameters	Designation	Level-1	Level-2	Level-3
Pulse Frequency (Hz)	A	12	14	16
Pulse Width(ms)	B	7	8	9
Welding speed(mm/sec)	C	1.0	1.5	2.0
Fictitious	D	d_1	d_2	d_3

(b) Simulated performance indicators

Test Run	Levels of LBW parameters				Welding Responses			
	A	B	C	D	α(mm)	β(mm)	γ(mm)	δ(K)
1	1	1	1	1	1.466	1.539	2.214	2475
2	1	2	2	2	1.215	1.304	1.915	2482
3	1	3	3	3	1.461	1.292	1.505	2714
4	2	1	2	3	1.050	1.097	1.966	2260
5	2	2	3	1	1.072	1.092	1.862	2313
6	2	3	1	2	3.599	2.000	2.147	2566
7	3	1	3	2	0.872	0.903	1.928	2146
8	3	2	1	3	3.670	2.000	2.108	2409
9	3	3	2	1	2.904	2.000	1.986	2347

(c) Simulated weld bead profiles for the nine test runs

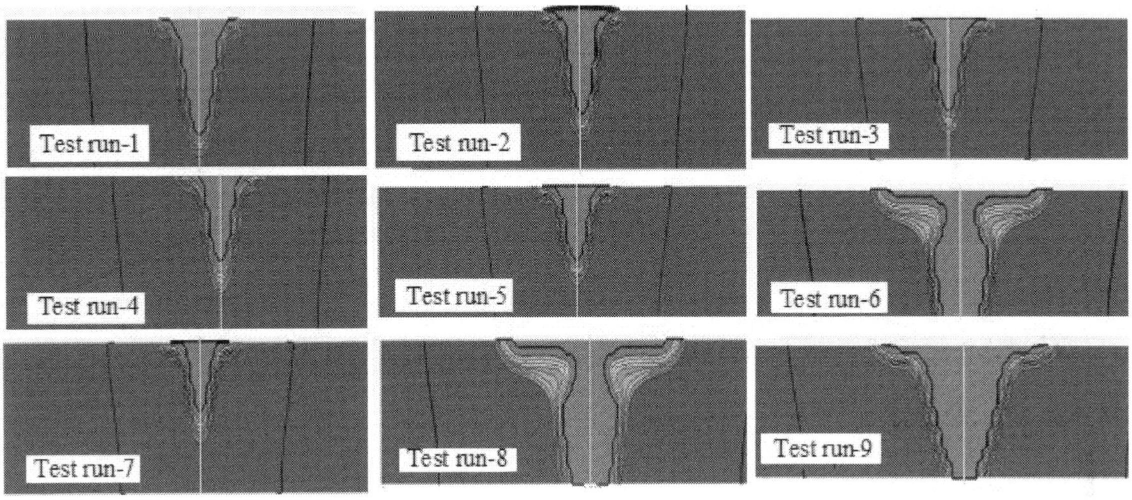
Test run-1
Test run-2
Test run-3
Test run-4
Test run-5
Test run-6
Test run-7
Test run-8
Test run-9

Table 2: Significance of LBW parameters through ANOVA

LBW parameters	1-Mean	2- Mean	3-Mean	Sum of Squares (SOS)	%Contribution
Width of Fusion zone, α(mm): grand mean=1.923 mm					
A	1.381	1.907	2.482	1.821	17.6
B	1.129	1.986	2.655	3.508	33.9
C	2.912	1.723	1.135	4.915	47.6
D	1.814	1.895	2.060	0.095	0.9
Depth of weld penetration, β(mm): grand mean = 1.470 mm					
A	1.378	1.396	1.634	0.123	8.1
B	1.180	1.465	1.764	0.512	33.9
C	1.846	1.467	1.096	0.845	56
D	1.544	1.402	1.463	0.030	2.0
Heat affected zone, γ(mm) :grand mean = 1.959 mm					
A	1.878	1.992	2.007	0.030	8.8
B	2.036	1.962	1.879	0.037	10.8
C	2.156	1.956	1.765	0.230	67.2
D	2.021	1.997	1.860	0.045	13.2
Maximum Temperature, δ (K) : grand mean = 2412 K					
A	2557	2379	2300	103395	44.6
B	2293	2401	2542	93308	40.3
C	2483	2363	2391	23790	10.3
D	2378	2398	2461	11190	4.8

Table-3: Estimates of the performance indicators.

Test Run	Parameters				Numerical Simulation	Eq.(17) (n_p=3)	RE (%)	Eq.(16) (n_p=4)	Range of estimates	
	A	B	C	D					From	To
Width of Fusion zone, α (mm)										
1	1	1	1	1	1.466	1.58	-7.8	1.466	1.466	1.712
2	1	2	2	2	1.215	1.24	-2.1	1.215	1.134	1.380
3	1	3	3	3	1.461	1.32	9.7	1.461	1.215	1.461
4	2	1	2	3	1.050	0.91	13.3	1.050	0.804	1.050
5	2	2	3	1	1.072	1.18	-10.1	1.072	1.072	1.318
6	2	3	1	2	3.599	3.63	-0.9	3.599	3.518	3.764
7	3	1	3	2	0.872	0.90	-3.2	0.872	0.791	1.037
8	3	2	1	3	3.670	3.53	3.8	3.670	3.424	3.670
9	3	3	2	1	2.904	3.01	-3.7	2.904	2.904	3.150
Depth of penetration, β (mm)										
1	1	1	1	1	1.539	1.47	4.5	1.539	1.398	1.539
2	1	2	2	2	1.304	1.37	-5.1	1.3040	1.304	1.445
3	1	3	3	3	1.292	1.30	-0.6	1.2920	1.232	1.373
4	2	1	2	3	1.097	1.10	-0.3	1.0970	1.037	1.178
5	2	2	3	1	1.092	1.02	6.6	1.0920	0.951	1.092
6	2	3	1	2	2.000	2.07	-3.5	2.0000	2.000	2.141
7	3	1	3	2	0.903	0.97	-7.4	0.9030	0.903	1.044
8	3	2	1	3	2.000	2.01	-0.5	2.0000	1.940	2.081
9	3	3	2	1	2.000	1.93	3.5	2.0000	1.859	2.000
Heat affected zone, γ (mm)										
1	1	1	1	1	2.214	2.15	2.9	2.214	2.053	2.214

2	1	2	2	2	1.915	1.88	1.8	1.915	1.778	1.939
3	1	3	3	3	1.505	1.60	-6.3	1.505	1.505	1.666
4	2	1	2	3	1.966	2.07	-5.3	1.966	1.966	2.127
5	2	2	3	1	1.862	1.80	3.3	1.862	1.701	1.862
6	2	3	1	2	2.147	2.11	1.7	2.147	2.010	2.171
7	3	1	3	2	1.928	1.89	2.0	1.928	1.791	1.952
8	3	2	1	3	2.108	2.21	-4.8	2.108	2.108	2.269
9	3	3	2	1	1.986	1.92	3.3	1.986	1.825	1.986
Maximum temperature, δ (K)										
1	1	1	1	1	2475	2509	-1.4	2475	2475	2557
2	1	2	2	2	2482	2496	-0.6	2482	2462	2545
3	1	3	3	3	2714	2665	1.8	2714	2631	2714
4	2	1	2	3	2260	2211	2.2	2260	2177	2260
5	2	2	3	1	2313	2347	-1.5	2313	2313	2395
6	2	3	1	2	2566	2580	-0.5	2566	2546	2629
7	3	1	3	2	2146	2160	-0.7	2146	2126	2209
8	3	2	1	3	2409	2360	2.0	2409	2326	2409
9	3	3	2	1	2347	2381	-1.4	2347	2347	2429

Table-4: Estimates of α for the 27 possible sets of LBW parameters.

Set Number	LBW parameters						Width of fusion zone, α (mm)		
	Levels			Pulse frequency, A (Hz)	Pulse width, B (ms)	Welding speed, C (mm/sec)	Eq. (18)	Min.	Max.
	A	B	C						
1	1	1	1	12	7	1.0	1.58	1.47	1.71
2	1	1	2	12	7	1.5	0.39	0.28	0.52
3	1	1	3	12	7	2.0	-0.20	-0.31	-0.06
4	1	2	1	12	8	1.0	2.43	2.32	2.57
5	1	2	2	12	8	1.5	1.24	1.13	1.38
6	1	2	3	12	8	2.0	0.65	0.55	0.79
7	1	3	1	12	9	1.0	3.10	2.99	3.24
8	1	3	2	12	9	1.5	1.91	1.80	2.05
9	1	3	3	12	9	2.0	1.32	1.21	1.46
10	2	1	1	14	7	1.0	2.10	1.99	2.24
11	2	1	2	14	7	1.5	0.91	0.80	1.05
12	2	1	3	14	7	2.0	0.32	0.22	0.46
13	2	2	1	14	8	1.0	2.96	2.85	3.09
14	2	2	2	14	8	1.5	1.77	1.66	1.91
15	2	2	3	14	8	2.0	1.18	1.07	1.32
16	2	3	1	14	9	1.0	3.63	3.52	3.76
17	2	3	2	14	9	1.5	2.44	2.33	2.58
18	2	3	3	14	9	2.0	1.85	1.74	1.99
19	3	1	1	16	7	1.0	2.68	2.57	2.81
20	3	1	2	16	7	1.5	1.49	1.38	1.62
21	3	1	3	16	7	2.0	0.90	0.79	1.04
22	3	2	1	16	8	1.0	3.53	3.42	3.67

23	3	2	2	16	8	1.5	2.34	2.24	2.48
24	3	2	3	16	8	2.0	1.76	1.65	1.89
25	3	3	1	16	9	1.0	4.20	4.09	4.34
26	3	3	2	16	9	1.5	3.01	2.90	3.15
27	3	3	3	16	9	2.0	2.43	2.32	2.56

Table-5: Estimates of β for the 27 possible sets of LBW parameters.

Set Number	LBW parameters						Depth of weld penetration, β (mm)		
	Levels			Pulse frequency, A (Hz)	Pulse width, B (ms)	Welding speed, C (mm/sec)	Eq. (19)	Min.	Max.
	A	B	C						
1	1	1	1	12	7	1.0	1.47	1.40	1.54
2	1	1	2	12	7	1.5	1.09	1.02	1.16
3	1	1	3	12	7	2.0	0.71	0.65	0.79
4	1	2	1	12	8	1.0	1.75	1.68	1.82
5	1	2	2	12	8	1.5	1.37	1.30	1.45
6	1	2	3	12	8	2.0	1.00	0.93	1.07
7	1	3	1	12	9	1.0	2.05	1.98	2.12
8	1	3	2	12	9	1.5	1.67	1.60	1.74
9	1	3	3	12	9	2.0	1.30	1.23	1.37
10	2	1	1	14	7	1.0	1.48	1.42	1.56
11	2	1	2	14	7	1.5	1.10	1.04	1.18
12	2	1	3	14	7	2.0	0.73	0.67	0.81
13	2	2	1	14	8	1.0	1.77	1.70	1.84
14	2	2	2	14	8	1.5	1.39	1.32	1.46
15	2	2	3	14	8	2.0	1.02	0.95	1.09
16	2	3	1	14	9	1.0	2.07	2.00	2.14
17	2	3	2	14	9	1.5	1.69	1.62	1.76
18	2	3	3	14	9	2.0	1.32	1.25	1.39

19	3	1	1	16	7	1.0	1.72	1.65	1.80
20	3	1	2	16	7	1.5	1.34	1.27	1.42
21	3	1	3	16	7	2.0	0.97	0.90	1.04
22	3	2	1	16	8	1.0	2.01	1.94	2.08
23	3	2	2	16	8	1.5	1.63	1.56	1.70
24	3	2	3	16	8	2.0	1.26	1.19	1.33
25	3	3	1	16	9	1.0	2.31	2.24	2.38
26	3	3	2	16	9	1.5	1.93	1.86	2.00
27	3	3	3	16	9	2.0	1.55	1.49	1.63

Table-6: Estimates of γ for the 27 possible sets of LBW parameters.

Set Number	LBW parameters						Heat affected Zone (mm)		
	Levels			Pulse frequency, A (Hz)	Pulse width, B (ms)	Welding speed, C (mm/sec)	Eq. (20)	Min.	Max.
	A	B	C						
1	1	1	1	12	7	1.0	2.15	2.05	2.21
2	1	1	2	12	7	1.5	1.95	1.85	2.01
3	1	1	3	12	7	2.0	1.76	1.66	1.82
4	1	2	1	12	8	1.0	2.08	1.98	2.14
5	1	2	2	12	8	1.5	1.88	1.78	1.94
6	1	2	3	12	8	2.0	1.69	1.59	1.75
7	1	3	1	12	9	1.0	2.00	1.90	2.06
8	1	3	2	12	9	1.5	1.80	1.70	1.86
9	1	3	3	12	9	2.0	1.60	1.51	1.67
10	2	1	1	14	7	1.0	2.27	2.17	2.33
11	2	1	2	14	7	1.5	2.07	1.97	2.13
12	2	1	3	14	7	2.0	1.87	1.78	1.94
13	2	2	1	14	8	1.0	2.19	2.09	2.25
14	2	2	2	14	8	1.5	1.99	1.89	2.05
15	2	2	3	14	8	2.0	1.80	1.70	1.86
16	2	3	1	14	9	1.0	2.11	2.01	2.17
17	2	3	2	14	9	1.5	1.91	1.81	1.97
18	2	3	3	14	9	2.0	1.72	1.62	1.78
19	3	1	1	16	7	1.0	2.28	2.18	2.34
20	3	1	2	16	7	1.5	2.08	1.98	2.14
21	3	1	3	16	7	2.0	1.89	1.79	1.95

22	3	2	1	16	8	1.0	2.21	2.11	2.27
23	3	2	2	16	8	1.5	2.01	1.91	2.07
24	3	2	3	16	8	2.0	1.82	1.72	1.88
25	3	3	1	16	9	1.0	2.13	2.03	2.19
26	3	3	2	16	9	1.5	1.92	1.83	1.99
27	3	3	3	16	9	2.0	1.73	1.63	1.80

Table-7: Estimates of δ for the 27 possible sets of LBW parameters.

Set Number	LBW parameters						Maximum temperature (K)		
	Levels			Pulse frequency, A (Hz)	Pulse width, B (ms)	Welding speed, C (mm/sec)	Eq. (21)	Min.	Max.
	A	B	C						
1	1	1	1	12	7	1.0	2509	2475	2558
2	1	1	2	12	7	1.5	2389	2355	2438
3	1	1	3	12	7	2.0	2417	2383	2466
4	1	2	1	12	8	1.0	2617	2583	2666
5	1	2	2	12	8	1.5	2496	2462	2545
6	1	2	3	12	8	2.0	2524	2490	2573
7	1	3	1	12	9	1.0	2758	2724	2807
8	1	3	2	12	9	1.5	2637	2603	2686
9	1	3	3	12	9	2.0	2665	2631	2714
10	2	1	1	14	7	1.0	2332	2298	2381
11	2	1	2	14	7	1.5	2211	2177	2260
12	2	1	3	14	7	2.0	2239	2205	2288
13	2	2	1	14	8	1.0	2439	2405	2488
14	2	2	2	14	8	1.5	2319	2285	2368
15	2	2	3	14	8	2.0	2347	2313	2396
16	2	3	1	14	9	1.0	2580	2546	2629
17	2	3	2	14	9	1.5	2460	2426	2509
18	2	3	3	14	9	2.0	2488	2454	2537
19	3	1	1	16	7	1.0	2253	2219	2302
20	3	1	2	16	7	1.5	2132	2098	2181

21	3	1	3	16	7	2.0	2160	2126	2209
22	3	2	1	16	8	1.0	2360	2326	2409
23	3	2	2	16	8	1.5	2240	2206	2289
24	3	2	3	16	8	2.0	2268	2234	2317
25	3	3	1	16	9	1.0	2501	2467	2550
26	3	3	2	16	9	1.5	2381	2347	2430
27	3	3	3	16	9	2.0	2409	2375	2458

Table-8: Single objective optimization function (ζ) with equal weighing factors (ω_1 and ω_2) for the simulated data of the width of fusion zone (α) and the depth of weld penetration (β) in Table-1.

Test Run	Levels of LBW parameters			Performance indicators		ζ_1	ζ_2	Single objective function
	A	B	C	α(mm)	β(mm)			
1	1	1	1	1.466	1.539	0.399	0.3	0.350
2	1	2	2	1.215	1.304	0.331	0.534	0.432
3	1	3	3	1.461	1.292	0.398	0.548	0.473
4	2	1	2	1.05	1.097	0.286	0.823	0.555
5	2	2	3	1.072	1.092	0.292	0.832	0.562
6	2	3	1	3.599	2	0.981	0	0.490
7	3	1	3	0.872	0.903	0.238	1.215	0.726
8	3	2	1	3.67	2	1	0	0.50
9	3	3	2	2.904	2	0.791	0	0.396

ANOVA (Analysis of Variance) on ζ			
Parameters	1-Mean	2-Mean	3-Mean
A	**0.418**	0.536	0.541
B	0.543	0.498	**0.453**
C	**0.447**	0.461	0.587

Table-9: Simulated weld bead profile for the optimal LBW welding parameters

Approach	Width of Fusion zone, α(mm)	Depth of weld, β(mm)	Size of HAZ, γ(mm)	Maximum temperature, δ (K)
Single Objective Optimization: Maximum Depth of weld ($A_3B_3C_1$)				
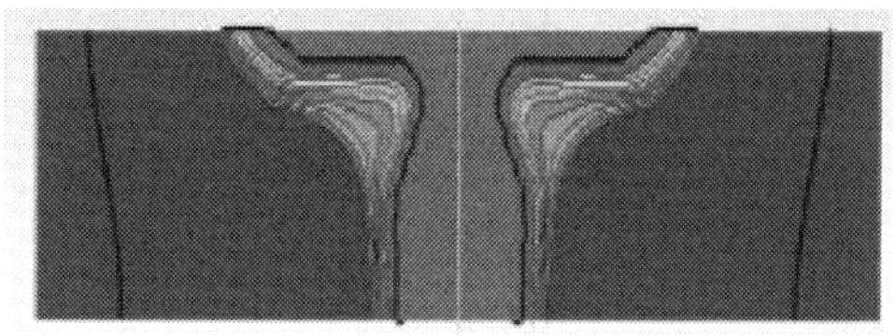				
Numerical simulations	4.92	>2	1.990	2490
Additive law	4.20	2.31	2.13	2501
Expected range	4.093-4.339	2.238-2.379	2.026-2.187	2467-2550
Multi-Objective Optimization: Optimum Depth and Width($A_1B_3C_1$)				
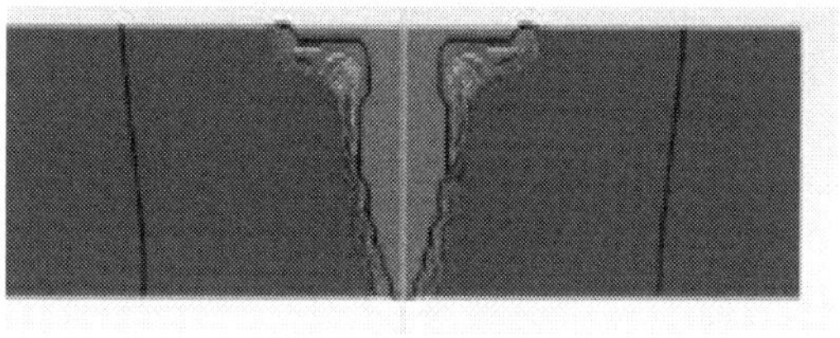				
Numerical simulations	3.166	>2	1.957	2884
Additive law	3.10	2.05	2.00	2757
Expected range	2.992-3.238	1.982-2.123	1.897-2.058	2723-2806

References

[1] Wang, S. H., Wei, M. D., & Tsay, L. W. (2003). Tensile properties of LBW welds in Ti–6Al–4V alloy at evaluated temperatures below 450 C. *Materials Letters*, *57*(12), 1815-1823.

[2] Casalino, G., Curcio, F., & Minutolo, F. M. C. (2005). Investigation on Ti6Al4V laser welding using statistical and Taguchi approaches. *Journal of materials processing technology*, *167*(2-3), 422-428.

[3] Kumar, G. S., Saravanan, S., Vetriselvan, R., & Raghukandan, K. (2018). Numerical and experimental studies on the effect of varied pulse energy in Nd:YAG laser welding of Monel 400 sheets. *Infrared Physics & Technology*, *93*, 184-191.

[4] Maio, L., Liberini, M., Campanella, D., Astarita, A., Esposito, S., Boccardi, S., & Meola, C. (2017). Infrared thermography for monitoring heat generation in a linear friction welding process of Ti6Al4V alloy. *Infrared Physics & Technology*, *81*, 325-338.

[5] Asséko, A. C. A., Cosson, B., Schmidt, F., Le Maoult, Y., Gilblas, R., & Lafranche, E. (2015). Laser transmission welding of composites–Part B: Experimental validation of numerical model. *Infrared Physics & Technology*, *73*, 304-311.

[6] Yang, Y., Gao, Z., & Cao, L. (2018). Identifying optimal process parameters in deep penetration laser welding by adopting Hierarchical-Kriging model. *Infrared Physics & Technology*, *92*, 443-453.

[7] Gale, W. F., & Totemeier, T. C. (Eds.). (2003). *Smithells metals reference book*. Elsevier.

[8] Cheepu, M., Muthupandi, V., & Che, W. S. (2018). Improving mechanical properties of dissimilar material friction welds. In *Applied Mechanics and Materials* (Vol. 877, pp. 157-162). Trans Tech Publications Ltd.

[9] Muralimohan, C. H., Muthupandi, V., & Sivaprasad, K. (2014). The influence of aluminium intermediate layer in dissimilar friction welds: Paper presented at "International Conference on Engineering Materials and Processes (ICEMAP 2013)", 23–24 May 2013, Tagore Engineering College, Chennai, Tamilnadu, India. *International Journal of Materials Research*, *105*(4), 350-357.

[10] Nancy Yanga, Milan Brandt and Shoujin Sun. 2009. Material science Forum online 04-17.

[11] Akman, E. R. H. A. N., Demir, A., Canel, T. I. M. U. R., & Sınmazçelik, T. (2009). Laser welding of Ti6Al4V titanium alloys. *Journal of materials processing technology*, *209*(8), 3705-3713.

[12] Mueller, S., Bratt, C., Mueller, P., Cuddy, J., & Shankar, K. (2008, October). Laser beam welding of titanium–a comparison of CO2 and fiber laser for potential aerospace applications. In *International Congress on Applications of Lasers & Electro-Optics* (Vol. 2008, No. 1, p. 1802). Laser Institute of America.

[13] Gao, X. L., Zhang, L. J., Liu, J., & Zhang, J. X. (2013). A comparative study of pulsed Nd:YAG laser welding and TIG welding of thin Ti6Al4V titanium alloy plate. *Materials Science and Engineering: A*, *559*, 14-21.

[14] Tzeng, Y. F. (2000). Process characterisation of pulsed Nd:YAG laser seam welding. *The International Journal of Advanced Manufacturing Technology*, *16*(1), 10-18.

[15] Torkamany, M. J., Hamedi, M. J., Malek, F., & Sabbaghzadeh, J. (2006). The effect of process parameters on keyhole welding with a 400 W Nd: YAG pulsed laser. *Journal of Physics D: Applied Physics*, *39*(21), 4563.

[16] Fang, X., Liu, H., & Zhang, J. (2014). Microstructure and mechanical properties of pulsed laser beam welded Ti-2Al-1.5 Mn titanium alloy joints. *Journal of materials engineering and performance*, *23*(6), 1973-1980.
[17] Fang, X., Liu, H., & Zhang, J. (2014). Microstructure and mechanical properties of pulsed laser beam welded Ti-2Al-1.5Mn titanium alloy joints. *Journal of materials engineering and performance*, *23*(6), 1973-1980.
[18] Kong, F., Ma, J., & Kovacevic, R. (2011). Numerical and experimental study of thermally induced residual stress in the hybrid laser–GMA welding process. *Journal of Materials Processing Technology*, *211*(6), 1102-1111.
[19] Frewin, M. R., & Scott, D. A. (1999). Finite element model of pulsed laser welding. *WELDING JOURNAL-NEW YORK-*, *78*, 15-s.
[20] Srinivasa Rao, B., Rudramoorthy, P., Srinivas, S., & Nageswara Rao, B. (2008). Effect of drilling induced damage on notched tensile strength and pin-bearing strength of woven GFR-epoxy composites. Mater. Sci. Eng. A. 472(1-2), 347–352.
[21] Singaravelu, J., Jeyakumar, D., & Nageswara Rao, B. (2009). Taguchi's approach for reliability and safety assessments in the stage separation process of a multistage launch vehicle. Reliability Engineering & System Safety. 94 (10), 1526–1541.
[22] Parameshwaran Pillai, T., Lakshminarayanan, P.R., & Nageswara Rao, B. (2011). Taguchi's approach to examine the effect of drilling induced damage on the notched tensile strength of woven GFR-epoxy composite. Adv. Compos. Mater. 20, 261–275.
[23] Sahiti, M., Raghavendra Reddy, M., Budi, J., Peter Praveen, J., & Nageswara Rao, B. (2016). Optimum WEDM process parameters of Incoloy®Alloy800 using Taguchi method. International Journal of Industrial and Manufacturing Systems Engineering. 1(3), 64–68.
[24] Bharathi, P., Priyanka, T.G.L., Srinivasa Rao, G., & Nageswara Rao, B. (2016). Optimum WEDM process parameters of SS304 using Taguchi method. International Journal of Industrial and Manufacturing Systems Engineering. 1(3), 69–72.
[25] Rajeev Kumar, D., Varma, P.S.S.K., & Nageswara Rao, B. (2017). Optimum drilling parameters of coir fiberreinforced polyester composites. American Journal of Mechanical and Industrial Engineering. 2(2), 92–97
[26] Konduri, S.S.S., Manideep Kumar Kalavala, V.D.S.R., Mandala, P., Manapragada, R.R., & Nageswara Rao, B. (2017). Application of Taguchi approach to seek optimum drilling parameters for woven fabric carbon fibre/ epoxy laminates. MAYFEB. J. Mech. Eng. 1, 29–37.
[27] Dutta, O.Y., & Nageswara Rao, B. (2018). Investigations on the performance of chevron type plate heat exchangers. Heat Mass Transf. 54(1), 227–239.
[28] Ross, P.J. (1989). Taguchi Techniques for Quality Engineering. McGraw-Hill, Singapore.
[29] Kumar, U., Gope, D. K., Srivastava, J. P., Chattopadhyaya, S., Das, A. K., & Krolczyk, G. (2018). Experimental and numerical assessment of temperature field and analysis of microstructure and mechanical properties of low power laser annealed welded joints. *Materials*, *11*(9), 1514.
[30] Jiang, D., Alsagri, A. S., Akbari, M., Afrand, M., & Alrobaian, A. A. (2019). Numerical and experimental studies on the effect of varied beam diameter, average power and pulse energy in Nd:YAG laser welding of Ti6Al4V. *Infrared Physics & Technology*, *101*, 180-188.
[31] Harish, M., Rao, S. S., Rao, B. N., Mahaboob, B., Praveen, J. P., & Reddy, A. I. (2021). Specific optimal AWJM process parameters for Ti-6Al-4V alloy employing the modified Taguchi approach. J. Math. Comput. Sci. 11(1), 292-311.
[32] Satyanarayana, G., Narayana, K. L., & Rao, B. N. (2018). Identification of optimum laser beam welding process parameters for E110 zirconium alloy butt joint based on Taguchi-CFD simulations. *Lasers in Manufacturing and Materials Processing*, *5*(2), 182-199.
[33] Satyanarayana, G. (2019). Thermal and fluid flow simulations in conduction mode laser beam welding of zirconium alloys. Ph.D. Thesis. Department of Mechanical Engineering, Koneru Lakshmaiah Education Foundation (Deemed to be University), Vaddeswaram, India. http://shodhganga.inflibnet.ac.in:8080/jspui/handle/10603/ 286801.
[34] Satyanarayana, G. Narayana, K.L., & Nageswara Rao, B. (2021). Incorporation of Taguchi approach with CFD simulations on laser welding of spacer grid fuel rod assembly, Materials Science and Engineering: B, Volume 269, 2021,115182
[35] https://doi.org/10.1016/j.mseb.2021.115182; https://www.sciencedirect.com/science/article/pii/S0921510721001422
[36]
[37] Kumar, C., Das, M., Paul, C. P., & Singh, B. (2017). Experimental investigation and metallographic characterization of fiber laser beam welding of Ti-6Al-4V alloy using response surface method. Optics and Lasers in Engineering. 95, 52-68.

[38] Rajulu, C. G., Krishna, A. G., & Babu Rao, T. (2018). An integrated evolutionary approach for simultaneous optimization of laser weld bead characteristics. Proceedings of the Institution of Mechanical Engineers, Part B: Journal of Engineering Manufacture. 232(8), 1407-1422.

[39] Caiazzo, F., Alfieri, V., Astarita, A., Squillace, A., & Barbieri, G. (2016). Investigation on laser welding of Ti-6Al-4V plates in corner joint. Advances in Mechanical Engineering. 9(1), 1687814016685546.

[40] Liu, J., Gao, X., & Zhang, J. (2016). Effect of heat input on the tensile damage evolution in pulsed laser welded Ti6Al4V titanium sheets. Journal of Materials Engineering and Performance. 25(11), 5109-5124.

[41] Cai, X., Sun, D., Li, H., Guo, H., Gu, X., & Zhao, Z. (2017). Microstructure characteristics and mechanical properties of laser-welded joint of γ-TiAl alloy with pure Ti filler metal. Optics & Laser Technology. 97, 242-247.

[42] Junaid, M., Khan, F. N., Rahman, K., & Baig, M. N. (2017). Effect of laser welding process on the microstructure, mechanical properties and residual stresses in Ti-5Al-2.5 Sn alloy. Optics & Laser Technology. 97, 405-419.

[43] Kashaev, N., Ventzke, V., Fomichev, V., Fomin, F., & Riekehr, S. (2016). Effect of Nd: YAG laser beam welding on weld morphology and mechanical properties of Ti–6Al–4V butt joints and T-joints. Optics and Lasers in Engineering. 86, 172-180.

[44] Hong, K. M., & Shin, Y. C. (2016). Analysis of microstructure and mechanical properties change in laser welding of Ti6Al4V with a multiphysics prediction model. Journal of Materials Processing Technology. 237, 420-429.

[45] Fan, Y., Chen, Z., Zhang, C. H., & Liu, A. M. (2015). A comparison of microstructure and mechanical properties of welded thin Ti6Al4V with three different types of laser. Materials Research Innovations,.19(sup4), S187-S192.

[46] Boccarusso, L., Arleo, G., Astarita, A., Bernardo, F., De Fazio, P., Durante, M. & Squillace, A. (2017). A new approach to study the influence of the weld bead morphology on the fatigue behaviour of Ti–6Al–4V laser beam-welded butt joints. The International Journal of Advanced Manufacturing Technology. 88(1-4), 75-88.

[47] Satyanarayana, G., Narayana, K. L., & Boggarapu, N. R. (2018). Numerical simulations on the laser spot welding of zirconium alloy endplate for nuclear fuel bundle assembly. *Lasers in Manufacturing and Materials Processing*, *5*(1), 53-70.

[48] Fluent, ANSYS 16 V User's Guide, ANSYS Inc.

[49] Voller V.R. & Prakash, C. (1987).A fixed grid numerical modelling methodology for convection-diffusion mushy region phase-change problems. International Journal of Heat and Mass transfer, 30 (8), 1709-1719. https://doi.org/10.1016/0017-9310(87)90317-6

[50] Brent, A.D., Voller, V.R. & Reid K.J. (1988). Enthalpy-porosity technique for modeling convection-diffusion phase change: Application to the melting of a pure metal. Numerical Heat transfer, 13 (3) 297-318DOI: 10.1080/10407788808913615

[51] Jalali, A. & Najafi, A.F. (2010).Numerical modeling of the solidification phase change in a pipe and evaluation of the effect of boundary conditions. Journal of Thermal Science, 19, 419-424. https://doi.org/10.1007/s11630-010-0403-z

[52] Rohit Rai (2008). Modeling of heat transfer and fluid flow in keyhole mode welding. Ph. D Thesis, Department of Materials Science and Engineering, The Pennsylvania State University (December 2008). https://modeling.matse.psu.edu//Theses_files/2008PhD_Rai.pdf

[53] Akbari, M., Saedodin, S., Toghraie, D., Razavjand, R. & Kowsari, F. (2014). Experimental and numerical investigation of temperature distribution and melt pool geometry during pulsed laser welding of Ti6Al4V alloy. Optics & Laser Technology. 59, 52-59. https://doi.org/10.1016/j.optlastec.2013.12.009.

[54] Singaravelu, J., Jeyakumar, D. & Nageswara Rao, B. (2012). Reliability and safety assessments on satellite separation process of a typical launch vehicle. Journal of Defense Modelling and Simulation. 9(4), 369-382.

[55] Rajyalakshmi, K. & Nageswara Rao, B. (2019). Expected range of the output response for the optimum input parameters utilizing the modified Taguchi approach. Multidiscipline Modeling in Materials and structures. 15(2), 508-522.

[56] https://doi.org/10.1108/MMMS-05-2018-0088

[57] Rajyalakshmi, K. & Nageswara Rao, B. (2019). Modified Taguchi approach to trace the optimum GMAW process parameters on weld dilution for ST-37 steel plates. ASTM International Journal of Testing and Evaluation. 47 (4), 3209-3223.

[58] Satyanarayana, G., Narayana, K.L., & Nageswara Rao, B. (2019).Optimal laser welding process parameters and expected weld bead profile for P92 steel. SN Applied Sciences (2019) 1:1291 | https://doi.org/10.1007/s42452-019-1333-3

[59] Dharmendra, B.V., Shyam Prasad Kodali, & Nageswara Rao, B. (2019). A simple and reliable Taguchi approach for multi-objective optimization to identify optimal process parameters in nano-powder-mixed electrical discharge machining of INCONEL800 with copper electrode. HELIYON. Vol.5 (2019) e02326
[60] http://doi.org/10.1016/j.heliyon.2019.e02326
[61] Dharmendra, B.V., Shyam Prasad Kodali& Nageswara Rao, B. (2019). Multi-objective optimization for optimum abrasive water jet machining process parameters of Inconel718 adopting the Taguchi approach. Multidiscipline Modeling in Materials and Structures.16 (2), 306-321.https://doi.org/10.1108/MMMS-10-2018-0175
[62] Tanya Buddi, Swadesh Kumar Singh & Nageswara Rao, B. (2018). Optimum Process Parameters for Plywood Manufacturing using Soya Meal Adhesive. Materials Today: Proceedings, 5, 18739-18744.
[63] Mooli Harish, Shanmukha Prasad, V., Sreekara Reddy, M.B.S., Koora Rajanikanth & Nageswara Rao, B. (2019). Optimal Process Parameters to Achieve Maximum Tensile Load Bearing Capacity of Laser Weld Thin Galvanized Steel Sheets. International Journal of Recent Technology and Engineering (IJRTE), 8 (4), 11682-11687.
[64] Shanmukha Prasad, V., Mooli Harish, Sreekara Reddy, M.B.S., Koora Rajanikanth, & Nageswara Rao, B. (2019). Optimal FSW Process Parameters to Improve the Strength of Dissimilar AA6061-T6 to Cu Welds with Zn Interlayer. International Journal of Recent Technology and Engineering (IJRTE), 8 (4), 11688-11695.
[65] Slavica Miladinovic, Sandra Velickovic, Kanulla Karthik, Devada Loknath & Boggarapu Nageswara Rao (2020). Optimal safe factor for surface durability of first central and satellite gear pair in Ravigneaux planetary gear set. TEST Engineering and Management. 83, 16504-16510.
[66] Mooli Harish, Rao, S.S. & Nageswara Rao, B. (2020). On machining of Ti-6Al-4V alloy and its parameters optimization using the modified Taguchi approach. TEST Engineering and Management. 83, 17007-17017.
[67] Slavica Miladinovic, Sandra Velickovic, Devada Loknath, Kanulla Karthik & Boggarapu Nageswara Rao (2020). Parameters identification and minimization of safety coefficient for surface durability of internal planetary gear using the modified Taguchi approach. TEST Engineering and Management, 83, 25108-25116.

Made in United States
Orlando, FL
04 November 2026

90200294R00024